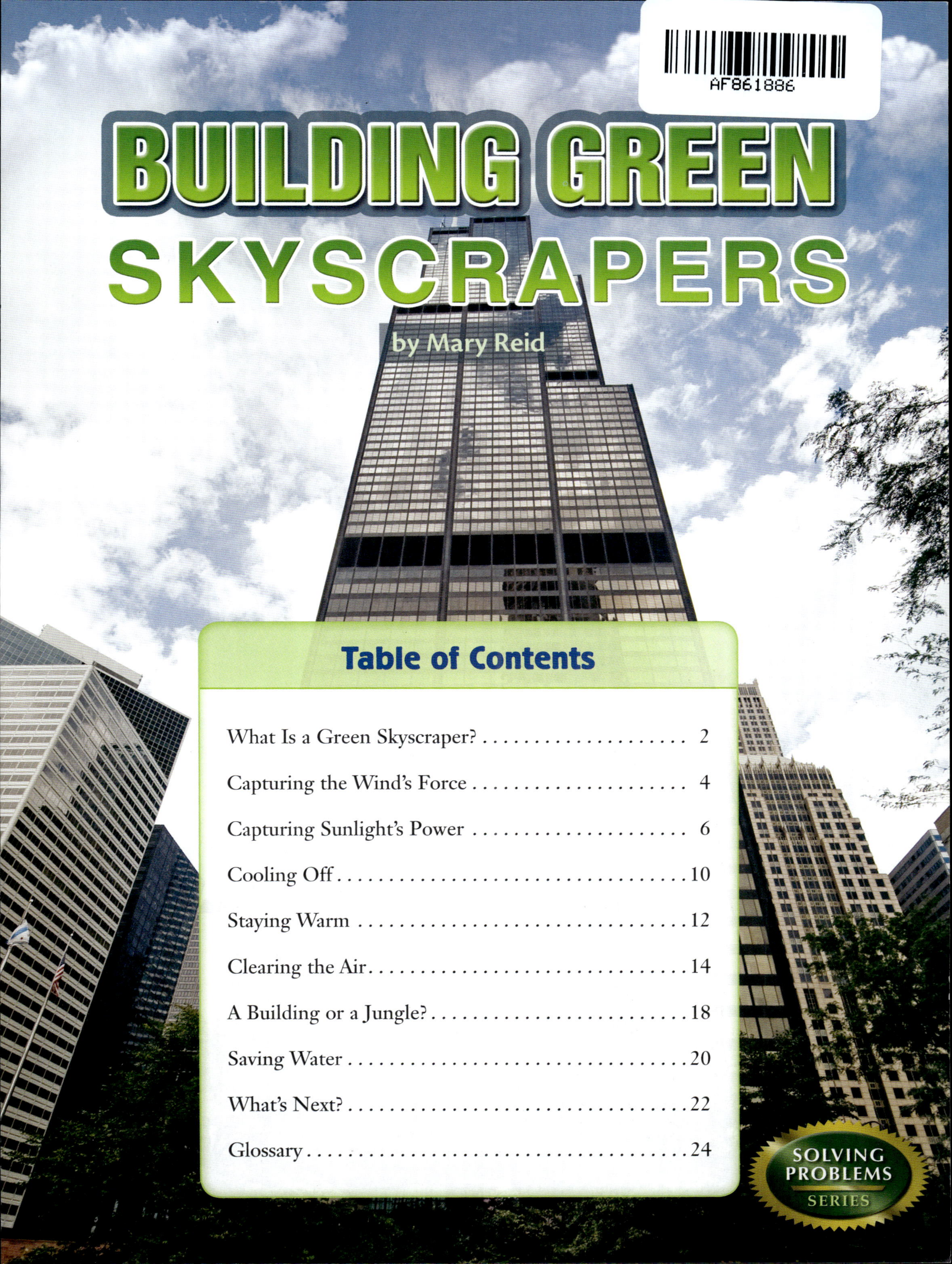

BUILDING GREEN SKYSCRAPERS

by Mary Reid

Table of Contents

SOLVING PROBLEMS SERIES

What Is a Green Skyscraper?

People need skyscrapers. They hold hundreds of apartments and offices on a small piece of land. Instead of spreading homes, office buildings, and parking lots over acres of forest or prairie, skyscrapers leave much more of the natural world untouched. They also make it possible for people to live or work close together.

There's a major problem with these cloud-skimming towers, however. Because they are so huge, they use an enormous amount of power for light, heating, and cooling. They give off a huge amount of **carbon dioxide**, a gas that **pollutes** the air, and they use incredible quantities of water.

Ordinary skyscrapers are not very good for the planet, and many of them are wasteful. Besides using too much energy and polluting, they also cost much more money to operate than they should. The old skyscraper model just doesn't make sense anymore.

Recently, scientists, **architects**, and builders have been working to create a new kind of skyscraper that saves power and water and helps keep the air clean. This cleaner construction is called "green building." The goal is to create buildings that will be pleasant, healthy places for people to live and work.

Architects have come up with some surprising ways to create green skyscrapers. From the looks of their designs, it seems they're having fun, too!

Going for the Gold—or a ZEB

A worldwide building contest is taking place. Green builders are creating the greenest buildings possible, and a group of judges give the buildings a rating: silver, gold, or the highest level, platinum.

Even better than a platinum rating, however, would be to build a **ZEB**, a Zero Energy Building. ZEBs produce the power they need by using sunlight, wind, or other sources of energy that don't pollute the earth. It is very hard to build a ZEB, and there are only a few of them in the world, including in China, France, and the United States. To be the builder of a ZEB is the highest honor of all.

The architect of this apartment building in Dubai wants every floor to turn slowly, independently of the others. If this building ever gets built, turbines between the floors will capture energy from the wind and turn it into electricity to power the entire building.

Capturing the Wind's Force

One way green builders make skyscrapers more efficient is by using clean energy, or energy that does not create pollution. It turns out that people don't have to burn millions of tons of coal or barrels of oil to produce electricity. And buildings don't have to puff huge clouds of carbon dioxide into the air. In fact, clean electric power can be produced right in the buildings where people live and work.

Wind power is one free source of clean electricity. Money is needed for the machinery that catches the wind, but nobody has to worry about running out of wind. Wind power does not produce harmful gases, either. That's why it's a main source of energy for green builders.

It's not an accident that the Bahrain [bah-RAYN] World Trade Center towers look like sails. The sail shapes guide the wind into the huge wind turbines between the towers.

How a Wind Turbine Works

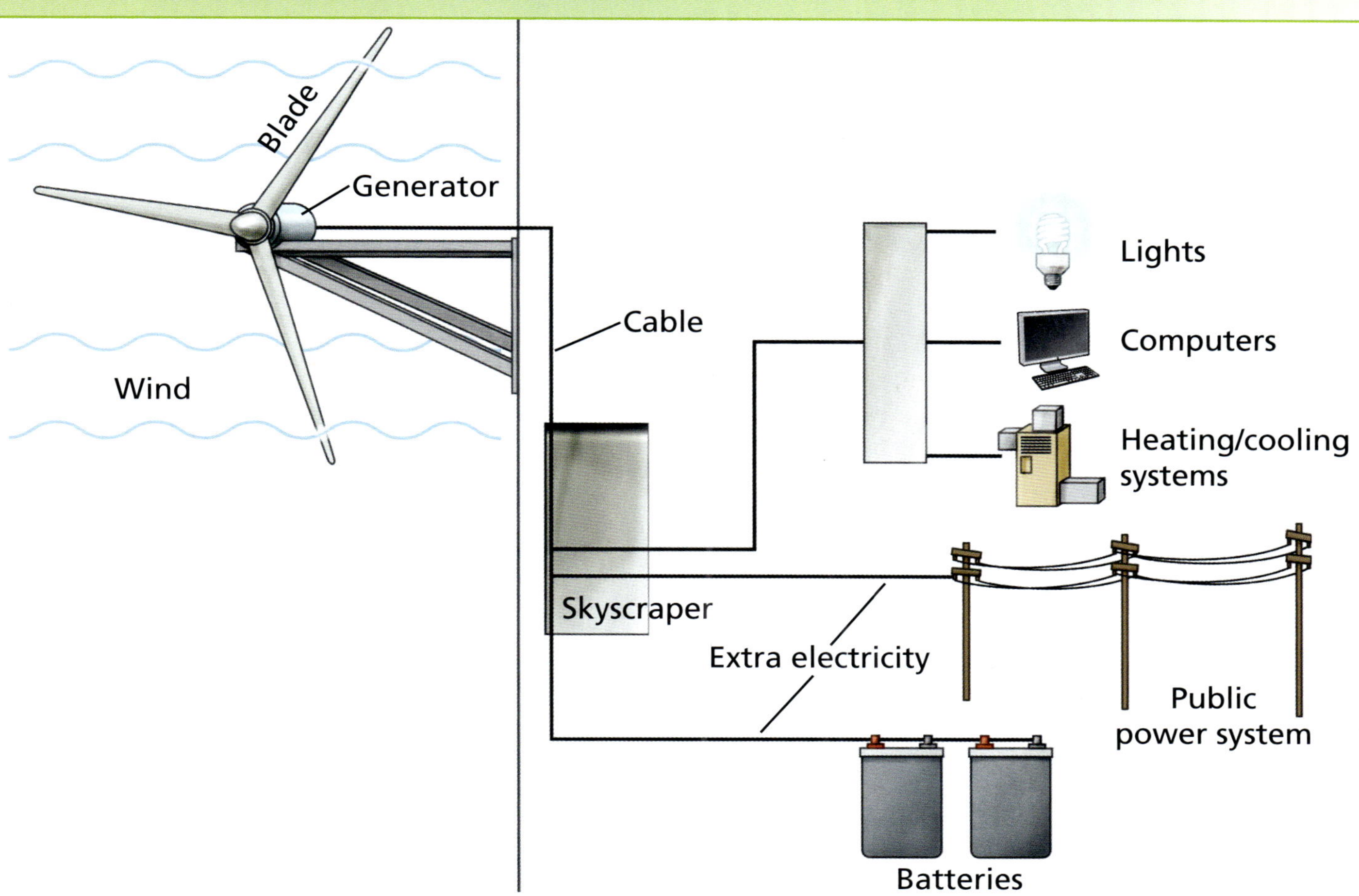

Many of the new skyscrapers use wind power. It works like this:

1. The wind spins the huge blades of a **wind turbine**.
2. The turbine runs a machine called a **generator** that changes the wind's force into electricity.
3. The electricity runs along a **cable** and into the skyscraper.
4. Newly generated electricity is used to light offices, power computers, and help heat or cool a building.
5. Extra electricity can be stored in batteries for later, or the building can sell it to the local power company for other people in the area to use.

Capturing Sunlight's Power

Like wind, sunlight is free, and it can also be used to generate electricity. It's no surprise then that green builders often use **solar panels** in their buildings to capture the sun's rays. For example, the Elithis [eh-LEE-tus] Tower in Dijon [dee-ZHON], France, has 330 solar panels on its roof.

The Elithis is greener than many bigger, fancier buildings. It's a real ZEB. In fact, it is more than a Zero Energy Building; it actually produces more energy than it uses.

The Elithis Tower doesn't just save energy. The ten-story office building also collects rainwater to use in its toilets.

How Solar Panels Work

Solar panels on the roof convert the energy from sunlight into electricity. The energy flows through electrical cables and can be used right away, stored in batteries, or sold to the local power company.

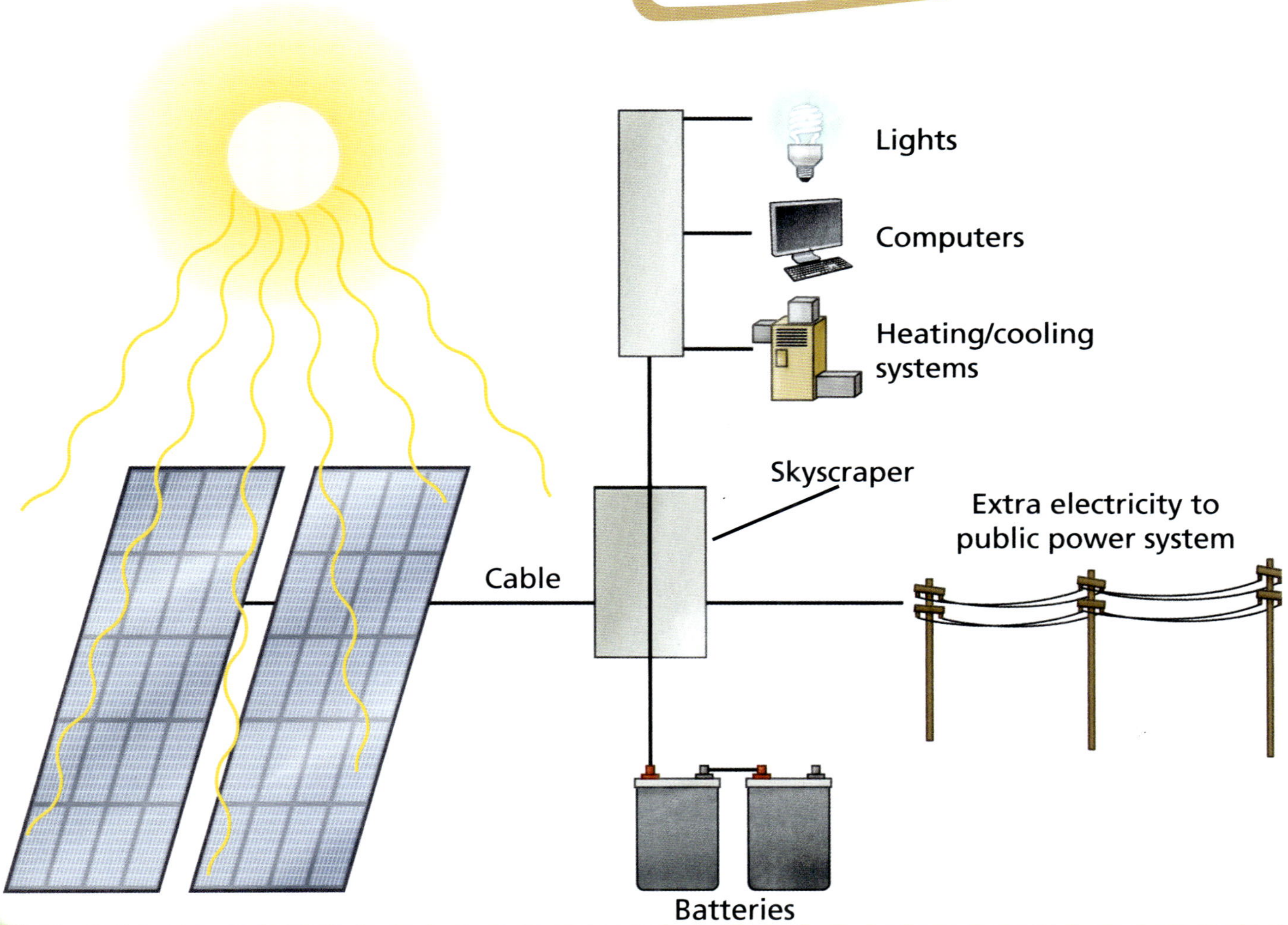

Saving Energy at the Elithis

Not only does the Elithis Tower produce its own solar power, the building also uses much less energy than most buildings. For one thing, there are almost no light switches in the Elithis. **Sensors** check the level of light in each room and whether a person is present. If the sun is shining brightly or there is no one in the room, the lights won't turn on.

In addition to solar panels, builders have started collecting the sun's energy in a new way—with **photovoltaic** [FO-toh-vohl-TAY-ik] windows. These may sound like windows with superpowers, and that's not far from the truth. Architects for the Willis Tower in Chicago, U.S.A., are testing a window that has tiny photovoltaic cells, or solar cells, right inside the glass. You can still see through the window, and it still lets in light. But the photovoltaic cells also help keep the room's temperature steady *and* they generate electricity. If the test windows do well, the architects at Willis Tower may put photovoltaic windows in most of the building.

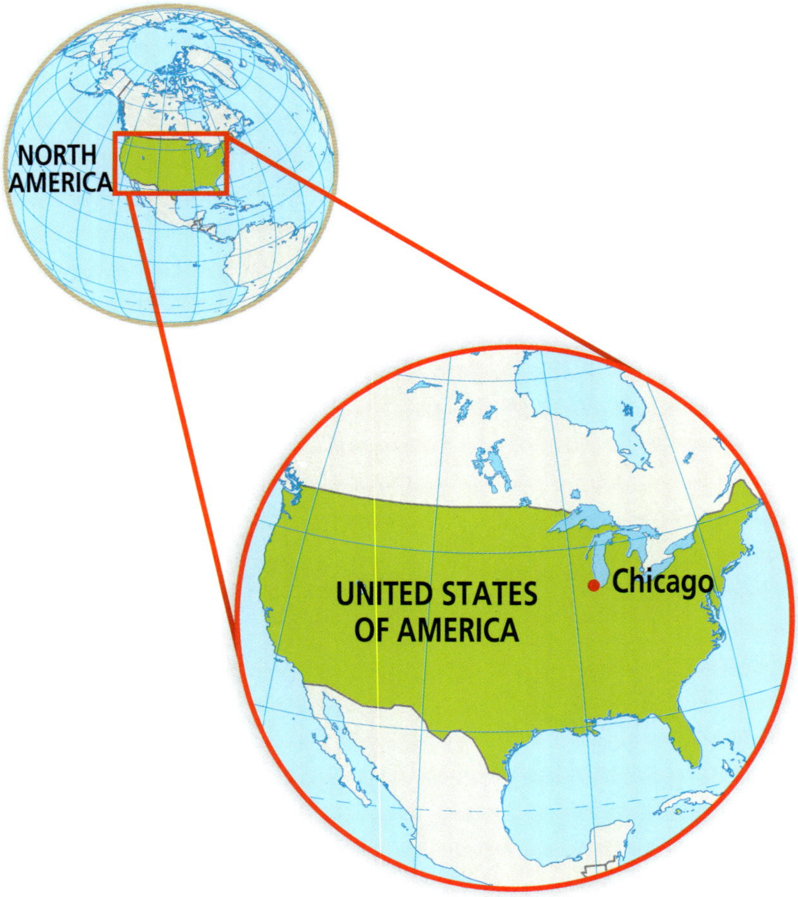

There isn't a lot of space for solar panels on top of a skyscraper. The Willis Tower is testing windows that can capture solar energy.

Cooling Off

Sunlight can be a great source of power, but its heat can also be a big problem, especially in a building with lots of windows. Even in winter, a building gets a lot of heat from sunlight. The many people, lights, and machines inside a building also give off heat. Ordinary skyscrapers have to be air-conditioned even in winter.

To help save energy and maintain comfortable inside temperatures, architects have started to use a new outside surface for their buildings: two layers of glass called a double glass skin. A chemical coating on the outside glass layer keeps out harmful rays of sunlight and too much heat. A gas between the two layers of glass keeps out both heat and cold. Fresh air from outside comes in through separate **vents** and moves through the building.

The Pearl River Tower in Guangzhou [gwahng-JOH], China, has a double glass skin. It is one of the greenest buildings in the world.

Staying Warm

Skyscrapers may use air-conditioning all year long, but some areas of the building, such as the lobby, might need heat in the winter. Heating and cooling buildings takes a great deal of **fuel** and electrical energy. This means a single skyscraper can give off a huge amount of pollution in the form of carbon dioxide. To reduce the pollution caused by artificial heating, green builders heat some buildings with the sun's warmth.

Solar heating is simple, and everyone has experienced it in a small way. If you touch a sunny place on a table, for example, it may feel warm. That's because materials like wood store heat. Solar heating doesn't take fancy equipment—all you need are the right windows, walls, and floors. You need a building that is designed to save energy.

How Solar Heating Works

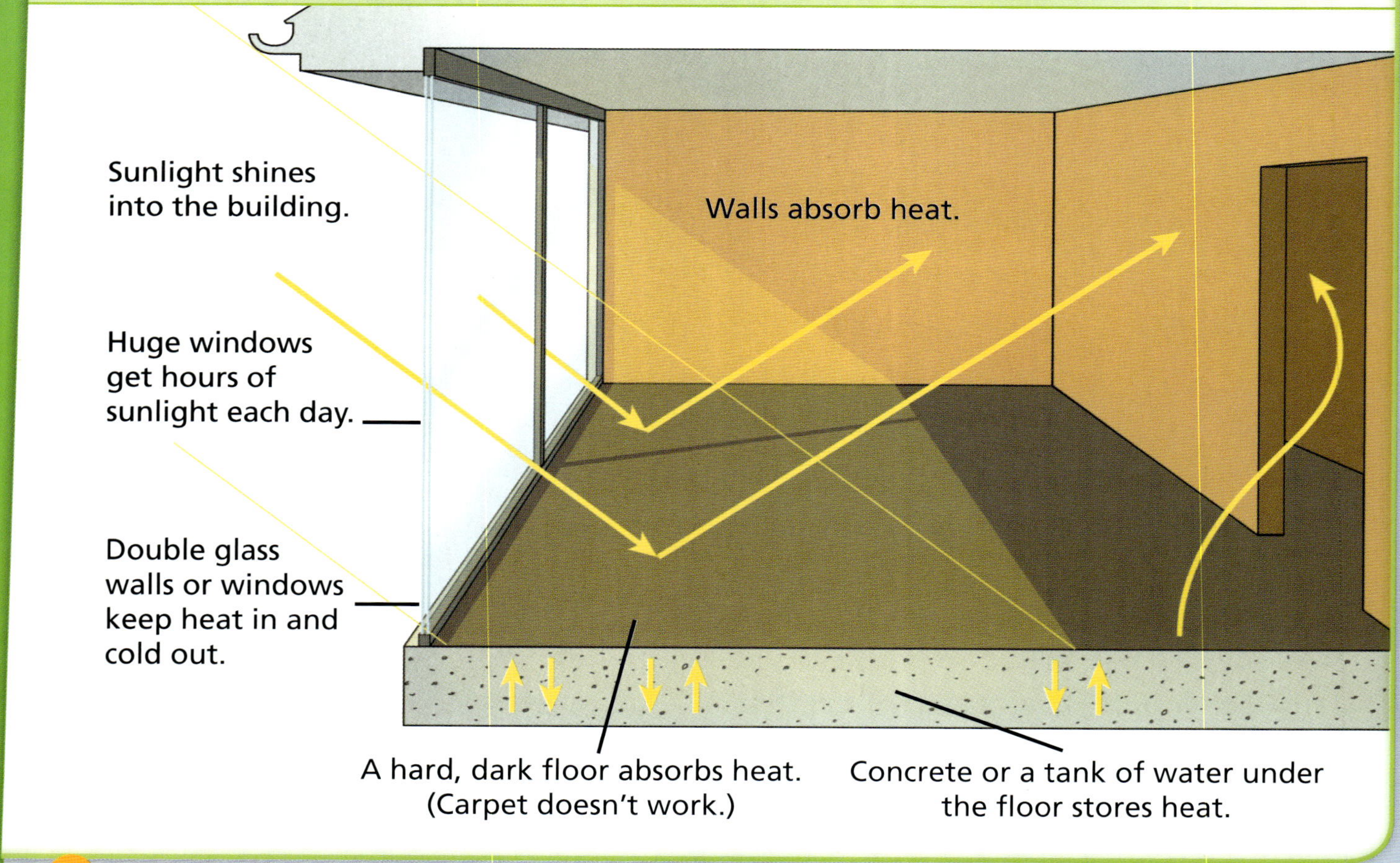

The Gherkin in London, United Kingdom, uses only half the power of other buildings its size. The skyscraper's six dark-colored stripes mark the location of open spaces running from the top of the building to the bottom. These shafts pull out hot air in the summer and keep warm air in during the winter, cutting down on air-conditioning and heating.

The skyscraper's double-glass skin also holds a few secrets. In the summer, blinds inside the glass close when sunlight heats up the building too much. In the winter, the glass allows solar heating to warm up cold spaces.

Clearing the Air

Green builders are also working to improve how skyscrapers smell. Many materials used in buildings give off harmful gases that smell like chemicals. Carpeting, ceiling tiles, drapes, furniture, and even the paint on walls may give off harmful gases that can make people sick. These gases are known as VOCs (short for volatile organic compounds). Some workers even have to quit their jobs because the VOCs in their building give them a serious illness.

VOCs are also found in car and bus fumes, but the amount of VOCs inside buildings is usually *ten times* what's in outside air. Sometimes it's much higher.

Unlike traditional builders, green builders use only materials that are healthy for humans. For example, there are now low-VOC paints that they can choose.

VOCs aren't the only problem with the air inside skyscrapers, however. Usually, windows don't open, so machines are needed to pull in fresh air and filter it throughout the building. If that system isn't working at its best, the air can become stuffy and stale, and people feel drowsy, get headaches, and can't concentrate.

Airplanes often have cleaner air than office buildings. The air inside a plane is completely refreshed with filtered air from outside about twenty times per hour. The air in an average office building is completely refreshed only twelve times an hour.

Scientists know that people work much better with lots of fresh air. That's why green builders include vents in their buildings that take in plenty of outside air. They also make sure their buildings have windows that people can open and close.

The Commerzbank [koh-MERTZ-bank] in Frankfurt, Germany, has several gardens in the building.

The builders of the Bank of America building in New York City created something really unusual. Their building takes in outside air and cleans it so carefully that even the used air is still unusually pure when it leaves the skyscraper. The building cleans air for the whole community around it!

The Bank of America building doesn't just offer a breath of fresh air. It also saves on air-conditioning thanks to forty-four giant tanks of water in the basement. Every night, the water is frozen, and during the day, the cool air coming off the melting ice is blown upwards into the office spaces.

A Building or a Jungle?

Many green buildings have rooftop gardens or planted areas within the buildings. These green areas help keep the inside air fresh, as well as give a feeling of being close to nature.

The ACROS Fukuoka in Fukuoka, Japan, is one example of a green and leafy skyscraper. The back of the building looks like a staircase, with each step covered in plants. This landscaped roof helps keep the inside air cool, even in the summer. Besides being unexpected and attractive, the green plants take in carbon dioxide and give off oxygen, which freshens the outside air.

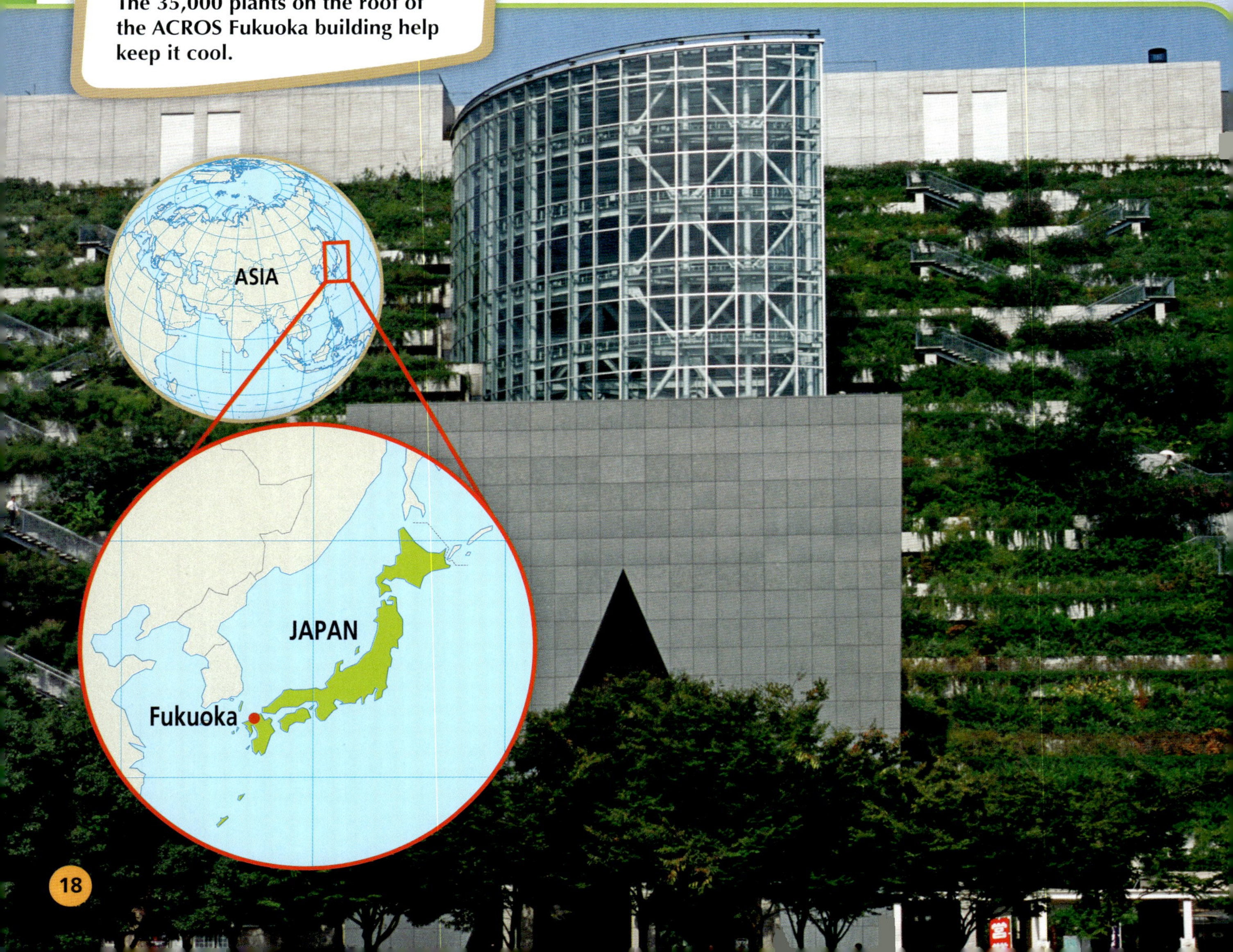

The 35,000 plants on the roof of the ACROS Fukuoka building help keep it cool.

The Bosco Verticale [BOS-koh vair-tee-KAH-lay] in Milan, Italy, has many trees on every floor. *Bosco verticale* means "vertical woods." Trees are planted on all the balconies. Here, too, the trees and plants clear carbon dioxide from the air around the building.

The Bosco Verticale looks like a forest rising into the sky.

Saving Water

One of the most important ways a building can help people and the planet is by saving water. At the Hearst Tower in New York City, the soothing sound of its famous waterfall sculpture echoes in the main **atrium**. The waterfall cools and adds moisture and freshness to the dry indoor air. The water comes from rainwater that the Hearst Tower collects on the roof. Rainwater is used in the cooling system and to water plants, too.

At the Hearst Tower in New York City, a waterfall runs along both sides of the escalator. It reuses rainwater.

Saving Steel

Much of the steel used to build the Hearst Tower was recycled. Green builders try to use recycled materials in everything they build. The diamond pattern used in the Hearst Tower is a new way to build that uses less steel than usual.

What's Next?

From buildings with green balconies and waterfalls, it isn't a great leap to urban farms. Designers are working (seriously!) on buildings where people could grow vegetables and other crops both inside and out.

Another idea is to create a surface for the walls of green buildings that would suck up the carbon dioxide in the air. Scientists are working on the idea right now.

Builders are also working to find or create building materials that are **nontoxic** and less wasteful than the concrete and steel they use now. One idea is to add something to mud bricks that would make them strong enough for skyscrapers.

Some people still want to build the "tallest building," so skyscrapers keep going up. But there's a new trend now—many builders care more about building the *greenest* building of all. Building design competitions around the world are receiving more and more plans for skyscrapers that use less energy, save water, and provide a pleasant place for people to live, work, and visit. It may only be a matter of time before one of them goes up in a city near you.

This photo illustration shows the Xero [ZEE-ro] Project, one of the winners of a design competition in Dallas, Texas, U.S.A. Though this solar-powered apartment building won't be built, it features the latest ideas in green building: a community farm and fruit trees.

Glossary

architect (AHR-kuh-tekt) n., a person who plans or designs buildings

atrium (EY-tree-uhm) n., a large open space inside a building, often with a high ceiling and a skylight in the roof

cable (KEY-buhl) n., a bundle of wires that carry electricity

carbon dioxide (KAHR-buhn dahy-AHK-sahyd) n., the gas people breathe out. Too much carbon dioxide, which also comes from cars and buildings, is not good for people, animals, or the Earth.

fuel (FYOO-uhl) n., oil, coal, or natural gas that is burned for energy

generator (JEH-nuh-rey-tuhr) n., a machine that changes one kind of power, such as wind power, into electricity

nontoxic (nahn-TAHK-sik) adj., not harmful or poisonous to living things

photovoltaic (foh-toh-vahl-TEY-ik) adj., capable of transforming the energy from sunlight into electricity

pollute (puh-LOOT) v., make harmful for living things by adding chemicals or waste

sensor (SEN-sohr) n., a device that responds to a physical sensation, such as light, and sends a signal for action, such as closing blinds

solar heating (SOH-luhr HEE-ting) n., using the sun's energy to heat buildings

solar panel (SOH-luhr PA-nuhl) n., a flat surface holding many photovoltaic cells that transform sunlight into electricity

vent (VENT) n., an opening to let air into or out of a room or building

wind turbine (WIND TUHR-buhn) n., a machine that changes wind power into electrical energy

ZEB (ZEE EE BEE) n., Zero Energy Building